Animal Communication

Wild World

Animal Communication

by Dr. Jim Flegg with
Eric & David Hosking

Newington Press

FRONT COVER: AN ELEPHANT SEAL ROARS.
BACK COVER: A STARLING SINGS.
TITLE PAGE: THE STRIPES ON A ZEBRA'S FACE HELP OTHER ZEBRAS TO IDENTIFY IT.

First published in the United States in 1991 by
Newington Press
2 Old New Milford Road
Brookfield, Connecticut 06804

First published in Great Britain in 1991 by Belitha Press Limited

Cataloging-in-Publication Data
Flegg, Jim
Animal communication / Dr. Jim Flegg, Eric Hosking and David Hosking. Brookfield, CT, Newington Press, 1991.
32 p.; ill.:
Includes glossary and index.
ISBN 1-878137-23-9
1. Animal communication. 2. Animal intelligence. 3. Animals—Habits and behavior. 4. Animal behavior. I. Title. II. Hosking, Eric. III. Hosking, David.
Fle 591.59

Words in **bold** are explained in the glossary.

Radio studios can communicate with thousands of people at once.

Contents

A smiling human and chimpanzee.

Animal Communication

We humans think we have some pretty good ways of communicating with each other. The telephone lets us speak to people in distant countries as if they were in the same room as ourselves. Communications satellites out in space are used to bounce radio and television signals to all parts of the world.

We use what is called body language too. A frown, for example, tells everyone that you are annoyed, while a smile tells them that you are happy.

Our most complicated way of communicating with each other is speech. Just think how many different sounds go to make up the enormous number of words we use. Singing birds use almost as many sounds, and their hearing is better than ours. Several kinds of whales can communicate over great distances, without any electronic aids. All sorts of animals, from tiny beetles to deer, rely on sound communication over short distances.

The bright colors of some animals serve to attract a mate. With other animals, such as wasps and skunks, colors serve as danger signals. With still others, colors help the animals hide.

Communicating with each other is one of the most important things that all creatures do. It is no wonder that we and other animals have found so many ways of getting the message across.

The display of an Emperor of Germany bird of paradise may attract a female.

Being seen

The birds of paradise live in the jungles of the tropical island of New Guinea. The females are fairly dull, but the males are perhaps the most exotic and colorful of all the world's birds. Most have feathers of several different brilliant colors, often sparkling with reflections, like thousands of tiny mirrors. They have lots of long, slender feathers called **plumes,** and when the male shows off or displays to a female, he fluffs these out. When he shakes them, he looks like a waterfall made of feathers. He may swing upside down from a branch to do this, and his performance attracts other males who join in the display.

Most birds of paradise make strange noises while they display. Some sound like whirring machinery. The display and noise attract the nearby female birds of paradise. They watch quietly for some minutes before flying off with the male of their choice.

Widow birds live in groups of ten to twenty on the grassy plains of Africa. Each group contains several males and several females. The females are about the size of thrushes. They have fawn feathers with brown and black streaks that give them excellent **camouflage** as they sit on their nests, low down in tufts of grass. Only one of the males, the leader, will be in full breeding **plumage.** He will be the father of all the nestlings. The other males have dull brown plumage like the females. The bright colors of the top male, with his extraordinarily long tail, help him to attract a large group of females.

The bright colors also make him conspicuous, and the long tail slows him down in flight. Because of this, he stands a greater chance of being caught by a **bird of prey.** If this happens, another of the males will immediately take his place. Within a week, its drab feathers will be replaced by the bright colors and long tail of the leader of the group.

Only the top male long-tailed widow bird has an eye-catching long tail.

Before there was electric light, children used to collect glowworms (also called fireflies or lightning bugs) in jars, to use as reading lights. Such light was less dangerous than candles, which, if knocked over, could set fire to a house.

Glowworms are actually beetles. In most species, both males and females have wings, and both give off light. However, some females, as the one shown here, are wingless. Glowworms' bright, greenish-white glow is produced by a mix of chemicals (including luciferin and luciferase) in **glands** in their bodies. The females' glow attracts the males. The males can spot a female and fly to her from several yards (meters) away.

The glow of a wingless female glowworm attracts males.

Sea slugs are beautiful and colorful animals.

Unlike many birds and mammals, simple animals, such as sea slugs, don't have sharp eyesight and hearing. They have to rely on simple sense organs. They have no ears, so they cannot hear. Their eyes are so poorly developed that all they can make out are vague colors and shapes. The bright colors and patterns of sea slugs have an important purpose. To find a suitable mate, the sea slugs have to be able to identify their own kind. Their strikingly different patterns help them to do this. In addition, the colors help to **camouflage** the sea slugs against the background of their tropical reef home.

Being heard

Although bush crickets live in meadows full of deep grass, they are able to send messages to other bush crickets nearby. Bright colors would be dangerous because there are plenty of birds ready to eat bush crickets. Most bush crickets are well **camouflaged** and blend in with their surroundings. They communicate successfully by using sound. On warm summer days, meadows can be very noisy places as male bush crickets sing. Their song may attract a mate or warn other males off their **territory.** Bush crickets do not have a voice box. Instead, they make their rasping song by rubbing their front wings together.

Bush crickets, living in long grass, communicate by song.

Deathwatch beetles can cause a lot of damage in houses.

The grubs or **larvae** of deathwatch beetles live in dead trees, burrowing tunnels in the wood. Sometimes, these beetles get into houses, where they can cause a lot of damage to the woodwork. The larvae tunnel away for up to six years, eating the wood as they go. Then they make **cocoons** of sawdust at the ends of their tunnels. Safe inside, they change into adult beetles about one third inch (1 cm) long and dull black.

When the adult beetles crawl out of their tunnels to the outside world, they put on an odd performance. They bang their heads upward on wood. Their heads are shaped like those of miniature rhinoceroses, and they use their horns to bang with. The beetles' banging may attract a mate. The slow, tapping noise is loudest in the middle of the night and seems to come from nowhere. In the past, superstitious people thought that it foretold a death in the house.

The drumming noise woodpeckers make serves to attract a mate or to show other woodpeckers the edges of their piece of woodland **territory.** Until about fifty years ago, some people thought that woodpeckers drummed by using their voices, just as we can imitate a drill by putting our tongues up behind our front teeth and saying *drrrrr. . . .* Other people thought that they made the noise by hammering rapidly with their beaks on a dry, dead branch. This was proved true by using microphones and slow-motion filming. You may wonder why woodpeckers do not get awful headaches by drumming. The answer is that between the beak and the rest of the skull there is a pad of spongy tissue, called **cartilage,** that acts like a shock absorber.

Woodpeckers make a drumming noise by hammering their beaks on dead wood.

Only male frogs and toads use their voices much. During the spring, their croaking can be heard around many ponds and ditches. Their calls attract females or warn of danger. The croaking is made by sucking air in and then blowing it out over some folds, called the vocal cords, in the **larynx.** In the same way, we make much more complicated sounds when we talk or sing. Air is blown over our vocal cords, or voice box. Frogs do not have big lungs, so they pass the same air backward and forward between the lungs and big pouches under or at the sides of their jaws. These vocal pouches swell and shrink as the frog croaks. When swollen tight, the pouches act like drums, which makes the croak sound even louder.

Even tiny frogs make a lot of noise when they call to each other.

For many years, sailors have known that whales could sing. The larger whales, such as humpbacks and blue whales, make groans, moans, and grunts, while smaller ones, such as porpoises and dolphins, make pings, whistles, and clicks. Dolphins use these sounds to find their way around. Whales often sing to keep in touch with each other. Some sounds, particularly low notes, travel better under water than in air. The humpbacks' calls are very efficiently produced, but they are not deafening, even if you are close to them. They can be heard by other humpbacks several hundred miles away. Submarines often copy the whales' method, using very low-frequency radio waves to keep in touch.

Humpback whales often sing to each other.

Male nightingales sing to the females arriving from Africa.

Nightingales spend the winter in Africa, flying back to Europe to breed in the summer. The males make this long **migration** first, a few days before the females. They return each year to the same patch of dense woodland. The 4,300-mile (7,000-km) journey is flown after dark, the birds perhaps navigating by the stars, and takes many nights. During the day, they rest and feed, gaining strength for the next stage of their journey.

Although they also sing during the day, nightingales are famous for producing their beautiful song at night. Once they have arrived, the males sing to attract a mate. This happens particularly on calm, clear nights, because that is when the females arrive at the end of their journey from Africa. Possibly the same female as last year will recognize a male's song.

Family ties

In a crowded sea lion colony, the first few days after the pups are born are the noisiest of the year. Most animals that live in colonies produce their young within a few days of each other. In these crowded conditions, it is vitally important that each mother sea lion can find her pup quickly when she comes to feed it. If she doesn't, it could starve. For the first few hours after the birth, mother and pup call to each other nonstop. The mother howls softly, and the pup cries back, sounding just like a human baby. In this way, they get to know each other's voices. The mother can then find her baby quickly by recognizing its hunger cries.

A mother sea lion nurses her pup.

A red deer stag roars a challenge to other males.

Who's boss?

For most of the year, red deer are shy and silent. Each winter, the male deer, called stags or bucks, lose their antlers and then grow a new set in the summer. Every year, the stags' antlers are bigger than the year before. When the stags are fully grown, with antlers 3 feet (1 m) or more across, they try to gather together a **herd** of female deer, or hinds. The stags compete with each other to keep the females. They have roaring contests, bellowing loudly for long periods to try and frighten each other off. They also try to impress younger deer with their huge antlers. Sometimes two closely matched stags fight, charging at each other and crashing their antlers together. Usually, the weaker stag will give up before any real damage is done, but occasionally the fight may be to the death.

Many animals are able to tell others what sort of mood they are in by the way they hold their bodies. Here, the stronger hyena is making sure that the other knows that it is going to have first bite at the **prey** that has been killed. It sends this message by raising the fur on its neck and shoulders (called its hackles) so that it bristles. This makes the hyena look as large as possible. The hyena may also snarl. The other recognizes who is boss and cowers down, with its head held low, trying to look as small as possible. The weaker hyena may also whine and curl up its lips in a sort of smile. It may even roll over on its back, to make sure that the stronger one knows that it has given in without a fight.

The hyena on the right is showing the other who's boss.

Look, don't touch!

Black and white are the colors that skunks use to signal a warning. Skunks are almost as big as badgers and have bushy tails like squirrels. They are normally peaceful animals, but if danger threatens, skunks have one of nature's most powerful weapons. Danger is most likely to be in the form of a **predator.** When a skunk sees one coming, it turns its tail toward the predator and fluffs it up. This makes the black and white stripes really obvious. Usually, the predator will take notice of this warning and turn away. If it doesn't, the skunk uses special **glands** under its tail to squirt out a cloud of droplets of awful-smelling liquid. This is so unpleasant that it will put off the hungriest hunter.

A skunk spreads its tail in warning.

A peacock butterfly shows the eyespots on its wings.

The peacock butterfly relies on the colors of its wings for defense. With its wings closed, the colors that can be seen on the underside are brown blotches. Then the butterfly looks like an old, tattered leaf. Its **camouflage** is superb as long as the butterfly stays quite still. If danger, such as a hungry bird, comes really close, then the butterfly suddenly flicks its wings wide open. This shows the bird the brightly colored spots, like eyes, on the butterfly's back wings. This usually startles the bird, and it flies off, leaving the butterfly alone. If it continues its attack, the bird bites at the eyespots, not at the butterfly's head or body. The butterfly will often escape with only a hole in its wing.

In the jungles of South America, several sorts of frogs have their very brightly colored skins that serve as a warning that they are unpleasant to eat or that they are poisonous. The birds, snakes, and mammals that eat frogs quickly learn from experience which frogs taste particularly nasty. The frogs' bright colors help the **predators** remember which of the frogs are to be avoided. Some South American Indian tribes use the poison from the frogs on their arrow tips.

The bright colors of the poison arrow frog warn that it will taste terrible.

The cuckoo's egg in this meadow pipit's nest is the one on the left.

The art of *not* communicating

European cuckoos are called nest **parasites** because the female cuckoos lay their eggs in the nests of other birds. The cuckoos leave the other birds to hatch the eggs and bring up the cuckoo chicks. Meadow pipits are birds often chosen to act as unwitting foster parents. A female cuckoo will keep a watch on a meadow pipit nest. When the meadow pipit has laid two or three of her own eggs, the cuckoo sneaks in, removes one of the meadow pipit's eggs, and lays one of her own in its place. The returning meadow pipit cannot spot the cuckoo's egg because its color and markings are such an excellent imitation of her own.

At first sight the zebras' black-and-white-striped coats make them stand out. You would think that stripes are not the best camouflage for animals on the African plains. But look closer at this photograph of a zebra **herd**. First, each zebra has a slightly different face pattern, so each can be recognized by other zebras. Second, when the herd is grazing safely, the

stripes are all lined up and the zebras are conspicuous, so no zebra loses sight of the herd. But should danger, such as a hunting lion, threaten, the zebras gallop off in all directions. The stripes that were once so clear become jumbled up and confusing. This dazzles the lion's eyes, so that it finds it difficult to single out a target zebra and make a kill.

A herd of zebras at a watering hole.

The young of garden birds such as thrushes, blackbirds, and robins hatch from their eggs tiny and naked, blind and helpless. They stay in their nest until they are fully grown and have all their feathers and are able to fly. Young birds like the woodcock are quite different. They hatch with their eyes open and with their legs strong and ready for running. They are dry a couple of hours after they have hatched. They can leave the nest and scamper about looking for food, under the watchful eyes of their parents.

Young birds that leave the nest immediately after hatching are covered in soft, warm, down feathers. They are very obedient. When the parents give a warning call, they crouch down and keep absolutely still. Then the patchwork colors of the down work like the **camouflage** markings on an army combat jacket, making them very difficult to see.

A young woodcock's coloring is perfect for camouflage.

Can you spot the ringed plover's eggs in this picture?

Ringed plovers make their nests by digging out a shallow hollow in a sandy bank. Most nest on beaches, above the high tide mark, but some breed away from the coast, in sand pits. At first sight, the nest seems to be just a simple saucer-shaped hollow, holding the four eggs. A closer look shows that the plovers have brought in small bits of shell and dry seaweed and put them around their nest. Against this background, the sandy-colored eggshells with small black marks are almost invisible. Hunting crows and gulls cannot find them. Ringed plovers face one main danger on holiday beaches. Sometimes, people don't see their eggs in time, and accidentally step on them.

Glossary

bird of prey a bird that hunts and kills other animals for food

camouflage colors or patterns that blend into the surroundings making something difficult to see

cartilage a tough flexible material usually fixed to the joints of bones; it helps bones to move smoothly

cocoon the case that a larva spins around itself for protection while it changes into an adult insect

glands organs that produce special chemicals for use by the body

herd a group of large animals, such as cattle, that feed and live together

larva (plural **larvae**) an insect after it has hatched from the egg; larvae have soft worm-like bodies

larynx part of the air passage in the throat; often called the voice box

migration the usually regular journeys that some animals make from one area to another, either in search of food or to breed

parasite a plant or animal that lives in or on another plant or animal and takes nourishment directly from it, or that depends on another plant or animal

plumage the feathers of a bird

plumes large or conspicuous ornamental feathers

predator an animal that lives by hunting other animals for food

prey an animal that is hunted by another animal

territory the area where an animal lives that it will defend against others

Further Reading

Animal Communication by Jacci Cole
(Greenhaven, 1988)
Beetles by Sylvia A. Johnson
(Lerner, 1982)
Butterfly and Moth by Paul Whalley
(Knopf, 1988)
Discovering Bees and Wasps by Christopher O'Toole
(Watts, 1986)
Discovering Deer by Jill Bailey
(Watts, 1988)
Fireflies by Sylvia A. Johnson
(Lerner, 1986)
Frogs and Toads by Jane Dallinger & Sylvia A. Johnson
(Lerner, 1982)
The Hyena by Judy Lockwood (ed.)
(Crestwood House, 1988)
Jane Goodall's Animal World: Chimps by Jane Goodall
(Macmillan, 1989)
A Natural History of Zebras by Dorcas MacClintock
(Scribner, 1976)
Seals and Sea Lions (Wildlife Education, 1985)
The Striped Skunk by Carl R. Green and William R. Sanford
(Crestwood House, 1987)
Whales by Donald Crump
(National Geographics, 1990)
Whales by Lionel Bender
(Watts, 1988)
The Zebra by Carl R. Green and William R. Sanford
(Crestwood House, 1988)

Index